Pass the Virginia Pharmacy Law Exam

A Study Guide for the FSDLE

First Edition

Alexis Long, Pharm D.

Pass the Virginia Pharmacy Law Exam – A Study Guide for the FSDLE

Copyright © 2009 Alexis Long

ISBN 978-0-557-07250-7

Published by Lulu

To purchase additional copies of this book, visit:

http://www.lulu.com/content/paperback-book/pass-the-virginia-pharmacy-law-exam/6156510

Introduction

Are you ready to become Virginia's next licensed pharmacist?

This book is designed to speed you on your way to licensure. It's arranged in roughly the same order as the study guide published by the Virginia Board of Pharmacy. The codes and statutes that the study guide references are translated into plain English, in question-and-answer format.

To get the most from this book, quiz yourself!

It's important to train your brain in active recall. To do this, use an index card or piece of paper to cover up the answers. Read the question, and formulate an answer on your own. Only reveal the answer after you've thought carefully about the question.

The study guide provided by the Board focuses heavily on state law. You probably know most of the relevant federal law already if you have spent internship hours working in a pharmacy, or are licensed as a pharmacist in another state.

Memorizing brand and generic drug names is less important for this exam than it is for the NAPLEX. You will need to be able to recognize if a drug is controlled or not. For example, you need to know the difference between Dilantin, Dilaudid, and diltiazem. But you will not be quizzed extensively on which drugs are in which schedules. If you feel shaky in this area, spend a few minutes on Appendix A, which lists the drug names used on the exam, along with the memory tricks I developed while studying. But spend most of your time on the law – this is a law exam, after all, not a drug exam.

Finally, be sure to allow yourself adequate study time. You will probably need to review this book several times before you can remember all of the material. However, it's well worth spending a few extra hours studying if it means you pass the exam on the first try.

Grab an index card and good luck! The sooner you pass the law exam, the sooner you can launch your career as a pharmacist in the state of Virginia.

Table of Contents

I. Licensing, Registration, and Inspection (24%)

Congratulations, you've got your license! When do you need to renew it?
Before December 31st each year. If you're newly licensed in October, November, or December, you don't need to renew the license until December of next year.

How many interns may a pharmacist supervise?
1 pharmacist supervises 1 intern.

How many hours of continuing education (CE) do you need to renew a license?
15 hours of CE. You can skip this for the first renewal. Extra hours don't carry over to the next year.

How long do you keep records of your CE?
Two years. You maintain these, not your employer.

Which types of programs count for CE?
- Approved by the ACPE (American Council on Pharmacy Education)
- Category 1 Continuing Medical Education (CME) on a pharmacy topic
- Approved by the Board of Pharmacy

How many hours of practical experience are required in VA?
1500 hours (300 outside school) of practical experience in the United States.

What if you renew your license late?
Within the first year – pay a renewal fee, plus a late fee, and complete all CE. You are still responsible for the same amount of CE as if you'd renewed on time.

After the first year, you must first apply for reinstatement, then renew.

What if you would like to get a VA license after having had an inactive license for more than one year?

Apply for reinstatement, pay a renewal fee, complete all CE. You need CE for all of the inactive years (up to 60 hrs CE).

What if you would like to get a VA license after having had an inactive license for more than *five* years?

You need proof of practice in another state or 160 hours as VA intern in the previous 6 months, and you must retake the VA law exam.

Does CE earned for a license in another state count toward your VA CE?

Yes.

Can you get an exemption from the CE requirement?

Yes, but it is rare and only granted for circumstances out of your control.

You can get a 1-year extension with a written request, and further extensions only with good reason. You are still responsible for all of the CE required for that time period.

If your license is inactive, are you still responsible for the CE?

Yes.

What dispensing activities can only be done by pharmacists, and not pharmacy technicians?

- Review prescriptions
- Take an oral prescription
- Conduct prospective drug review
- Counsel patients and provide drug information
- Communicate with prescriber or agent about any change to therapy except a refill
- Verify a prescription prior to dispensing
- Supervise pharmacy interns and technicians

Can a pharmacy intern do these things?

Yes, if a pharmacist directly supervises the intern.

When can a pharmacy technician at a nuclear pharmacy accept oral prescriptions?

A technician at a nuclear pharmacy can accept oral prescriptions for non-patient specific radiopharmaceuticals.

What is the maximum number of technicians a pharmacist can supervise?

Four. As many as two of them can be trainees who aren't registered technicians yet.

Can a technician compound extemporaneous preparations?

Yes, but a pharmacist must personally supervise the preparation.

What does a pharmacist have to do after an order has been prepared, prior to delivery?

Inspect and initial it to certify accuracy.

What is required to become certified as a pharmacy technician?

A technician must hold current certification by the Pharmacy Technician Certification Board (PTCB) OR complete an approved training program and pass an exam.

Can an unregistered technician work while training to get registered?

Yes, as long as they are directly supervised by a pharmacist.

Does a technician who only works in a free clinic have to be registered?

Yes. The Board lets them get a limited-use registration (free clinic only). The Board waives the initial registration fee and first examination fee for this. If they want to switch to an unlimited registration, all they have to do is pay the current renewal fee.

When does a technician need to renew his or her license?

Before December 31st each year. A technician who is newly licensed July 1st or later doesn't need to renew the license until December of next year.

What if a technician renews his or her license late?

- Within the first year, the technician must pay a late fee, a renewal fee, and show proof of CE.

- After 1 year, the technician pays a reinstatement fee, a renewal fee, and shows proof of CE.

- After 5 or more years, the technician will need to apply for recertification. The same requirements apply as getting certified for the first time.

How much CE is required each year for a registered technician?
5 hours of CE. It is possible to get an extension of 1 year with written request – but the technician still needs to do the full amount of CE.

How long is a technician required keep records of CE?
Two years. The technician maintains these, not the employer.

What documents must be maintained by a pharmacy that employs technicians?
- A site-specific training program and manual

- Documentation of completion of this training for each technician (kept for 2 years after they leave the job)

- For technician trainees, proof that they are enrolled in an approved training program

How long can a technician trainee work before getting certified?
9 months.

Define supervision – can a pharmacist supervise over the phone? Leave written instructions? Use any kind of electronic monitoring?
No. Supervision must be in person, by a pharmacist who is physically present and available for immediate oral communication.

What are grounds for revocation or suspension of a pharmacist's license?
- Negligence

- Unprofessional conduct

- Incompetence due to a mental or physical condition

- Use of drugs or alcohol to the extent that the pharmacist is rendered unsafe to practice

- Fraud or deceit

- Engaging in activities beyond the scope of the license
- Allowing unlicensed persons to practice duties for which a license is required
- Violating or cooperating in violating an law or regulation relating to pharmacy
- Revocation of registration by DEA or other federal agency
- Engaging in theft or diversion of controlled substances
- Violating any federal or state drug law
- Revocation of license in another state
- Conviction of any felony or misdemeanor involving moral turpitude
- Issuing statements intended to deceive or defraud about his professional service
- Has practiced in such a manner as to be a danger to the health and welfare of the public
- Has failed to comply with continuing education requirements

Who can suspend the license of a pharmacist?
Any health regulatory board.

Is a hearing required?
No. The license may be immediately suspended if there is danger to public health or safety. A hearing will be scheduled within a reasonable time.

What do you do if you know a pharmacist has done something that endangers public health?
Submit allegations in writing to the Board of Pharmacy.

If your license is revoked or suspended in another jurisdiction, or if you have been convicted of a felony or adjudicated incapacitated, what will the director of the Department of Health Professions do?
Immediately suspend the license without a hearing.

What if the conviction is not final?
The director can decline to suspend the license if there is a likelihood of injury to the public if the pharmacist's services aren't available.

If your license is suspended, how do you apply for reinstatement?

You can bring counsel and witnesses to the hearing. You need ¾ of the board members at the hearing to vote for reinstatement, and the Board might reinstate upon terms or conditions it deems appropriate.

Can you participate in "kickbacks," fee-splitting, or special charges in exchange for prescription orders?

Yes, but it must be fully disclosed in writing to the patient and any third party payor.

Can you interfere in the patient's right to choose a supplier of medication?

No.

Can you cooperate with anyone in denying a patient the opportunity to select his supplier of prescribed medications?

No.

When a permit to conduct a pharmacy is issued, who or what is it issued to?

The Pharmacist-In-Charge (PIC).

What has to be on the application to conduct a pharmacy?

- The pharmacy's corporate name and trade name
- Any pharmacists in addition to the PIC who will work there
- The hours the pharmacy will be open
- The pharmacy owner (if someone other than the pharmacist making the application)

What must be done if the hours of operation of the pharmacy change?

If the change will last more than a week, report the change to the Board in writing *and* post a notice 14 days in advance in a conspicuous place.

What happens if the PIC leaves the job?

The pharmacy permit must be surrendered to the Board, and an application for a new permit made.

In what circumstances does the PIC need to turn in the pharmacy permit and make a new application?

If there is any change in ownership of the pharmacy, such as a change in partnership composition or acquisition.

Can a pharmacy keep operating if it fails to designate a new PIC after the old one leaves?

No. It is not even allowed to maintain a stock of prescription drugs.

If the pharmacy does not have a PIC, the pharmacy receives a notice that it has no valid permit, and is required to do what?

Dispose properly of all Schedule II through VI drugs and devices on the premises within fifteen days.

What if the pharmacy fails to do this?

The Board of Pharmacy will seize all Schedule II through VI drugs and devices still on the premises, and notify the owner.

To claim the drugs, the owner must pay a fee for the storage cost, and provide for proper disposition of the drugs.

If the owner has not claimed the drugs after six months, the drugs may be destroyed.

What does a new PIC have to do when taking over responsibility for a pharmacy?

Inventory all Schedule I, II, III, IV and V drugs on the day he or she becomes PIC. The inventory must be completed *before* opening for business.

How long is a pharmacy permit good for?

One year.

Can you be PIC of more than one pharmacy?

Yes, you may be PIC of a maximum of 2 pharmacies.

Can the non-pharmacist owner of a pharmacy make decisions that override the decisions of the PIC or the pharmacist on duty?

No. That would be considered practicing pharmacy without a license.

When the PIC leaves, how long does the pharmacy have to file an application naming a new one?

14 days.

Is it possible to get a special or limited-use pharmacy permit?

Yes. This is up to the discretion of the Board of Pharmacy. You need to list the regulatory requirements you want waived and why. You must also maintain a policy and procedure manual outlining the type of operation and schedules of drugs maintained.

One pharmacy is being acquired by another. Do current patients need to be notified?

Only if prescription records are going to be used for something other than continuity of pharmacy services at substantially the same level. If so, written notice must be given 14 days prior to the acquisition.

If the pharmacy is changing locations, or if structural changes are being made, does the Board need to be notified?

Yes. The pharmacy needs to be inspected by an agent of the Board.

Give the Board at least 14 days notice before you would like the inspection. The dispensing area needs to comply with requirements at the time of the inspection. If re-inspection is required, there is a fee.

Before the inspection, can you stock drugs in the pharmacy?

No, and they can't be moved into a new location either. *If you don't have a permit, you can't stock drugs.*

Is the pharmacy required to display its permit?

Yes, it must be displayed prominently.

When is the pharmacy required to renew its permit?

Annually.

If the pharmacy is being sold, can the pharmacy permit be transferred to the new owners?

No, the permit is nontransferable.

Can the pharmacy operate with no pharmacist on duty?

No.

What is the PIC of a hospital pharmacy responsible for?

- Establishing procedures for and assuring maintenance of the proper storage, security, and dispensing of all drugs used throughout the hospital.

- Providing reviews of drug therapy

Is the PIC of a hospital pharmacy required to attend meetings of the Pharmacy & Therepeutics (P&T) committee?

No, but most of them do.

The hospital pharmacy is going to open a satellite pharmacy. Does the Board need to be notified?

Yes, an inspection is required. No drugs may be stocked until the inspection is completed and approval given.

The PIC can delegate ordering and distribution of certain drugs to non-pharmacy personnel. Which drugs?

- Large volume parenteral solutions that contain no active therapeutic drug other than electrolytes

- Irrigation solutions

- Contrast media

- Medical gases

- Sterile sealed surgical trays that may include a Schedule VI drug; an

- Blood components and derivatives, and synthetic blood components and products

How do these things have to be stored?

They must be locked when authorized staff is not present.

One medical gas must be locked up all the time, even when authorized staff is present. Which one?

Nitrous oxide.

What are the requirements for the prescription department?

- It must be at least 240 square feet (not including patient waiting area or the area used for counseling, devices, cosmetics, and proprietary medicines).

- Rest rooms and stock rooms must not be accessible through the prescription area. If there is a restroom within the prescription area, it must be for pharmacy personnel only.

- There must be a sink with hot & cold running water.

- There must be a refrigerator.

Can you use a trailer as a pharmacy?

No, a pharmacy must be permanent and secure.

What is the minimum equipment required in a pharmacy?

- A current dispensing information reference source

- Either an electronic scale *or* a set of Prescription Balances (sensitive to 15 milligrams) and weights

- Other equipment, supplies, and references consistent with the pharmacy's scope of practice and with public safety

Are pharmacies required to have a security system?

Yes, if the pharmacy ever closes. No security system is required for pharmacies that are open 24 hours a day, 7 days a week, and are always staffed by a pharmacist.

What are the requirements for devices used as pharmacy security systems?

- It must be a "generally accepted" device, such as sound, microwave, photoelectric, or ultrasonic

- It must be in working order and have a backup power supply

- It must fully protect the prescription area

Who may have access to the alarm system?

Pharmacists working at the pharmacy.

When is the security system activated?

Whenever the prescription department is closed for business.

What are the requirements for the enclosure around the prescription department?

- It must protect controlled drugs from unauthorized entry and theft

- It must be high enough to keep people from reaching in to gain access to drugs

- The gap between the floor and the door must be no more than 6 inches

- The door must be at least as high as the adjacent structure

- There must be a lock on the door

Who can have keys to the pharmacy?

Only pharmacists who practice at the pharmacy and are authorized by the PIC may have keys to the pharmacy, or to the alarm access code.

Is the PIC allowed to keep a key available for emergencies?

Yes, the key or alarm access code may be placed in a sealed envelope or other container with the pharmacist's signature across the seal. It must be kept in a safe or vault within the pharmacy or other secured place.

If the pharmacist is unavailable, can a technician go into the prescription department to get a prescription that has already been filled and checked?

Yes, but only in an emergency. An emergency includes an unplanned absence of a pharmacist scheduled to work during regular pharmacy hours, and an inability to obtain alternate pharmacist coverage.

The technician must be accompanied by a member of the pharmacy's management or administration.

Does the PIC need to give permission in order for someone to go into the prescription department in the absence of a pharmacist?

Yes, verbal permission is required.

After the technician goes into the prescription department without a pharmacist, what needs to happen?

- A record must be made and kept 1 year.

- The technician must reseal the key or alarm access code after using it.

- A pharmacist must change the access code within 48 hours.

Can prescriptions prepared for delivery to the patient be stored outside of the prescription department?

Yes, they may be placed in a secure place if access to the prescriptions is restricted to designated clerical assistants.

If prescriptions are delivered to a patient when a pharmacist is not present, do any records need to be kept?

Yes. A log must include the patient's name, prescription number(s), date of delivery, and the signature of the person receiving the prescription. This record must be kept for one year.

Can SII drugs be stored with the other schedules of drugs?

Yes, or they may be separately locked up.

Can SII drugs be left unlocked while there is a pharmacist on duty and prescription department is open?

Yes.

What about controlled paraphernalia?

They must be stored where the pharmacist can exercise "reasonable supervision and control."

Where are expired medications stored?

They must be kept separate from the other drugs.

Can drugs in the emergency room be left out in the open?

No. they must be kept where unauthorized people can't have access.

Can a doctor or other practitioner dispense drugs in the emergency room?

Yes, if it is a bona fide medical emergency, pharmaceutical services are not available, and it is permitted by hospital. The same labeling requirements apply as for prescriptions dispensed anywhere else. A separate record must be kept for 2 years.

Is after-hours access to a hospital pharmacy allowed?

Yes, if the PIC authorizes it, a nurse may get *emergency* medications that are in the manufacturer's original packaging, or in units prepared and labeled by a pharmacist.

Do any records need to be kept?

Yes, a record must be kept for 1 year with the date, patient's name, drug(s) taken, and the signature of the nurse.

More than one week

The pharmacy is going to close for a short while because the pharmacist/owner is on vacation. What are the rules?

If it will be closed for more than 1 week, the pharmacy must post a notice 60 days prior to closing *or* mail a notice 14 days prior to closing to every customer with refills available.

The notice must include the name of the pharmacy that records will be transferred to unless patients indicate preference to the contrary.

If there is a change in pharmacy ownership, which records must be transferred?

The last two years worth of prescription dispensing records and patient records.

The pharmacy refuses to process a request for prescription dispensing records or other records tendered in accordance with law. What is this?

This constitutes a closing.

The pharmacy is closing. Does the Board need to be notified?

Yes, you must report how you intend to dispose of all SII-VI drugs, prescription and dispensing records, including the name and address of licensee who will receive the drugs and records, and the date of transfer.

The pharmacy is going to change ownership. Does the Board need to be notified?

You must notify the Board 14 days prior to the change in ownership, and transfer two years worth of prescription and dispensing records.

The pharmacy's hours are going to change. Who needs to be notified?

No notice is required if the hours are expanding. NO NOTICE if expanding

If the hours are changing in any other way, you must notify the public 14 days in advance. 14 days for short.

Aside from pharmacy employees, who may be allowed access to copies of records of drug shipments?

- An agent of the Board of Pharmacy
- An agent of the Superintendent of the Department of State Police for the purpose of drug diversion investigations

Can they police copy and remove patient records?

Only if the records are relevant to a specific investigation.

17

What are the requirements if the police agent copies the records onto "magnetic storage media"?

The agent must leave a duplicate with the person who maintains the records (the pharmacist).

What are the requirements if the police agent takes the original?

The agent must leave a receipt and provide a copy to the person maintaining the records within a reasonable time afterwards.

↑ leave a receipt ↑ provide a copy for records

Is a police agent allowed to obtain samples of any stock of drugs?

Yes. They must give a receipt. *Receipt*

What records are off limits to the police agent?

Financial data, sales data other than shipment data, pricing data, personnel data or research data.

The Board of Pharmacy is responsible for what?

- Regulating the quality, integrity, safety and efficacy of drugs, cosmetics and devices
- Maintaining records
- Controlling factors contributing to abuse
- Promoting scientific or technical advances in the practice of pharmacy
- Regulating the impact of costs to the public

The Board has authority to implement a "pedigree system" to apply to certain schedules or certain drugs that are more subject to counterfeiting. What is a "pedigree"?

"Pedigree" means a paper document or electronic file recording each distribution of a controlled substance from sale by a pharmaceutical manufacturer through acquisition and sale by any wholesale distributor until sale to a pharmacy or other person dispensing or administering the controlled substance.

Can the Board collect samples of drugs, devices, and cosmetics for inspection?

Yes.

Which establishments may the Board enter and inspect?

A pharmacy, or any other place in Virginia where drugs, cosmetics or devices are manufactured, stored or dispensed.

When may an agent of the Board come to inspect?

During normal business hours.

If a pharmacy, manufacturer, or other entity has more than one place of business, how many DEA registrations do they need?

One for each place of business.

DEA for each

Is a warehouse considered a place of business that requires DEA registration?

No.

Is a marketing office where no drugs are stored a place of business that requires DEA registration?

No.

How often does a manufacturer need to re-register with the DEA?

Every year.

How often does a pharmacy need to re-register with the DEA?

Every 3 years.

* Can transfer DEA not
permit

If you are an agent of a registrant, do you need to be registered yourself?

No.

If a doctor has a DEA number, does this mean that he or she can prescribe controlled substances to treat patients addicted to narcotics?

No. Any doctor may use narcotics for pain control, but treating addiction requires special registration.

More information on laws regulating narcotic addiction treatment can be found in Section III – Reviewing Prescriptions.

Can a doctor working for a hospital dispense controlled substances under the hospital's DEA number?

Yes, if it is part of the physician's duties at the hospital. Also, the hospital has to keep a current list of internal codes and the corresponding individual practitioners.

Who else is exempt from the requirement to register with the DEA?

- Agents and employees of registrants (e.g., employees of a pharmacy)
- Practitioners affiliated with registrants (e.g., doctors working at a hospital)
- Law enforcement officials

A pharmacy wants to supply a long term care facility with an automated dispensing machine. What are the DEA registration requirements?

The pharmacy needs to get a separate registration for each facility. If another pharmacy also wants to put an automated dispensing machine at the site, the second pharmacy needs their own DEA number for that site as well.

What are the separate business activities that each require their own DEA number?

- Manufacturing
- Distributing
- Reverse distributing
- Dispensing
- Research (SI drugs)
- Research (SII-V drugs)
- Narcotic treatment (including compounder)
- Importing
- Exporting
- Chemical analysis

A patient has been prescribed Vicodin. Can she take it with her on her trip to the Bahamas? Or is this considered "exporting"?

If you have a lawful prescription for yourself or an animal, you may "import" or "export" it without DEA registration.

Are there any organizations that are exempt from paying DEA registration fees (but not exempt from getting the appropriate DEA registration)?

Yes. Facilities operated by a United States agency (including the U.S. Army, Navy, Marine Corps., Air Force, and Coast Guard), or by any state, are exempt from paying fees.

A facility wants to handle SI drugs for research purposes. The facility already has a DEA number. What else must they do?

Apply for a modified DEA registration by submitting 3 copies of the research protocol. There is no fee for modification, and it is handled just like a regular registration application.

When is a registrant's DEA registration considered terminated?

If the registrant dies, if the business ceases legal existence, or if the person or professional stops practice.

If a pharmacy ceases business, what are they responsible for doing?

- Notify the DEA
- Return the certificate of registration and unexecuted order forms
- Dispose of controlled substances properly

Can a pharmacy transfer its registration to the new owners after being acquired?

Yes. The registrant must submit a written request to the Deputy Assistant Administrator in Washington, DC 14 days before the proposed transfer.

14 days

Do you have to get any special paperwork from the Special Agent in Charge?

No. If the date the transfer was supposed to occur arrives, and you haven't heard that it can't occur, then the registration may be transferred.

What must be done on the date of transfer?

- Complete an inventory of all controlled substances
- Use an order form if transferring SI-II substances
- Registrants that are required to file reports (e.g., registrants that routinely destroy controlled substances) must file a final report

transfer must use 222 to move SII

List the ways to legally dispose of a controlled substance.

- Request assistance from the Special Agent in Charge
- Transfer to another registrant authorized to possess the substance
- Deliver the drugs to the DEA
- Destroy the drugs in the presence of a DEA agent or other authorized person

What are the requirements for registrants that routinely dispose of controlled substances?

The Special Agent can authorize routine destruction without prior approval every time. The registrant must keep records and file periodic reports.

The DEA would like to inspect a facility. What is required?

They need an administrative warrant (from a judge). *must have warrant*

They do not need a warrant if:

- The inspection is required prior to the DEA granting some kind of authorization
- The inspection is not constitutionally prohibited
- It is an emergency
- There is an imminent threat to public health and safety
- The owner consents in writing

What does the DEA need to show to receive an administrative warrant?

Administrative probable cause. This is different from criminal probable cause.

How often are manufacturers of SI-II drugs inspected?

Every year.

How often are distributers of SI drugs inspected?

Every year.

How often are all other registrants inspected?

As needed.

Which entities are required to register with the FDA?

Manufacturers and distributers are, pharmacies aren't.

pharmacies don't have to register w/ the FDA

If a business is registered with the FDA to manufacture drugs or Class II or III devices, how often is it inspected?

Every 2 years.

2 years

Can FDA inspect whenever they want?

Yes, if they think there is a reason to inspect.

What can't they inspect?

Financial data, sales data other than shipment data, pricing data, personnel data (other than data as to qualification of technical and professional personnel), and certain research data.

Does the FDA need to provide notice prior to inspection?

Yes, for each inspection (but not for each entry during an inspection).

Can non-financial records be inspected?

Yes. If you are required to keep the records, the FDA is authorized to inspect the records.

Who *cannot* be employed somewhere where they have access to controlled substances?

- Person convicted of a felony offense relating to controlled substances
- Person who had a DEA registration application denied or registration revoked or surrendered for cause

A theft of controlled substances has been discovered. How soon afterwards must the DEA be notified?

In writing, one day after discovery.

II. Ordering, Receiving, and Managing Drug Inventory (21%)

Who can a manufacturer or wholesaler sell SVI drugs/devices to?

Anybody who is authorized to administer, prescribe, or dispense the drug/device.

Who can a manufacturer or wholesaler sell SII-V drugs to?

- Another manufacturer/wholesaler

- A licensed practitioner: a pharmacist, pharmacy, MD, DO, podiatrist, dentist, or veterinarian

- An official (federal, state, territory, district, county, municipality) on government business

- A ship or aircraft captain, if there is no regular doctor on board, for the actual medical needs of people on board. The vessel cannot be in port.

- A person in a foreign country in compliance with relevant federal laws.

What is required to transfer SII drugs?

An official written order – the DEA 222 form.

How long must records of an SII transfer be kept?

Both the purchaser and receiver need to keep a copy for 2 years.

Can a pharmacy ever act as a wholesale distributor of small quantities of prescription drugs without being a licensed wholesaler?

Yes. Wholesale controlled substances sale can't exceed 5% of total controlled substance sales.

Wholesale oxygen sales can't exceed 5% of total oxygen sales

Do you need to use a DEA 222 form to transfer SII drugs from a central fill pharmacy to a retail pharmacy?

No.

Can a registrant delegate the ordering of SII drugs to someone else?

Yes, the registrant can give power of attorney to one or more agents.

Where do registrants get DEA 222 forms?

The forms are given to registrants by the DEA when they are first registered.

Subsequent forms are ordered using form 222a.

Order w/ 222a

Can you store or execute blank DEA forms at an address other than the one on the form?

Yes, but must be easily accessible if investigated by law enforcement.

Who fills out the DEA 222 form, the purchaser or the distributor?

The party receiving the drugs. The pharmacy fills out the form when purchasing drugs. However, if the pharmacy is returning drugs, the distributor must fill out the form.

Where do the three copies of DEA form 222 go when drugs are purchased?

The purchaser fills out the form, and keeps Copy 3.

Copies 1 & 2 are sent to the supplier.

After the order is filled, Copy 1 is kept by the supplier, and Copy 2 is sent to the DEA.

How long is a completed form 222 good for?   *222 for 60 days*

60 days from when it is executed.

How soon does Copy 2 need to get to the DEA?

By the close of the month in which the shipment is filled.

If filled in partial shipments, the close of the month in which the last shipment was sent or the 60-day validity period expires.

What are special rules for the Defense Supply Center of the Defense Logistics Agency?

They may get partial shipments at different times, to locations other than the location printed on the form 222, for a period of 6 months from date of the order. They must be buying drugs for delivery to armed services establishments within the US.

If a supplier can't fill all or part of an order, what can they do with the form 222?

They may endorse it to another supplier to fill. Only the original supplier can do this.

When is a supplier prohibited from filling an order written on a DEA form 222?

If the order is illegible or altered.

What happens if an order is refused? *Returned back to purchaser q reason*

The supplier returns form 222 (copies 1&2) to the purchaser with a statement of reason for refusal. *x all 3 copies to purchaser*

The purchaser retains all 3 copies plus the statement from the purchaser in their files.

The purchaser can't re-use the form.

What if a DEA Form 222 is lost in the mail?

The purchaser executes another, attaches a statement with the serial number and date of lost form, and a statement that the drugs weren't received.

The purchaser keeps Copy 3 of the lost form, plus Copy 3 of the new form, plus the statement.

The supplier gets Copies 1&2 of the new form and fills the order as usual.

If the supplier eventually gets the lost form, they must mark it "Not Accepted" and return it to the purchaser.

What if a DEA Form 222 is lost or stolen (but not lost in the mail)?

Report it to the special agent in charge of the DEA divisional office for that area.

If the form(s) are found later, that must be reported too.

Which copies of DEA Form 222 are the purchaser and supplier responsible for retaining? For how long?

Copy 1 – kept by the supplier for 2 years

Copy 2 – kept by DEA

Copy 3 – kept by the purchaser for 2 years

Unaccepted or defective forms – kept by purchaser for 2 years, along with statement of reason rejected

all 2 years including rejected

Where are the executed forms kept?

At the location printed on the form.

What's the exception to the rule that a pharmacist may only buy from a licensed distributor?

In an emergency, a pharmacy may purchase drugs from another pharmacy.

In what situations may you sell, purchase, or trade a drug sample?

Never.

Who (besides wholesalers) can sell SI-VI drugs?

On an emergency basis, drugs can be transferred (sold or borrowed) between pharmacies.

Drugs may also be transferred within health care entities such as hospitals that are under common control.

What's required for a drug company to give out samples?

A written request from a prescriber, and a written receipt of samples given. The prescriber may authorize the drugs to go to a hospital or pharmacy instead of his or her office.

The manufacturer must keep the records for 3 years.

Samples = 3 years.

When SI-V drugs are received, what does the receiver need to record?

- Date
- Name & address of supplier
- Kind & quantity of drug received
- If receiving coca leaves or crude opium, the amount of morphine, cocaine, or ecgonine that can be made from them

Does the pharmacy need to maintain SI and SII inventories and records separate from other records?

Yes.

all I & II need to be separate

Can SVI inventories and records be maintained with SIII-V inventories and records?

Yes. You can keep SVI records on their own or with SIII-V, but not with other pharmacy records.

When is it OK to store order forms, prescriptions, invoices, and inventories of SII-V records offsite?

If the DEA authorizes it and the records are retrievable within 48 hours.

In what order must the records be stored?

Chronologically.

How often does a pharmacy need to take an inventory of SI-V drugs?

Biennially (every 2 years). *Inventory MUST 2 years*

The pharmacy opened for business on January 1, and opening inventory was taken. There is a theft in December, nearly 2 years later, and an inventory was done. Does the pharmacy still have to take a biennial inventory a few weeks later in January?

Yes.

What is considered the opening inventory?

The most recent one, the inventory taken in December.

Who signs and dates the inventory?

The person taking the inventory. This may be a pharmacist or technician.

Can it be taken orally using a recording device?

Yes, if promptly reduced to writing.

If the pharmacy is open 24 hrs/day, the inventory must indicate what?

Whether the receipt or distribution of drugs on the inventory date occurred before or after the inventory was taken.

If the pharmacy is NOT open 24 hrs/day, the inventory must indicate what?

If it was taken at close of business or opening of business.

A new pharmacy opens for business. What are the inventory requirements?

There must be a complete inventory of SI-V drugs on the FIRST day of business. If there are no controlled substances on hand, record that as the inventory.

A pharmacy fails to do this for 3 days. How many separate offenses does this constitute?

Three.

When there is a change in PIC, what are the inventory requirements for the *old* PIC?

Complete an inventory of SI-V drugs, and return the pharmacy permit to the Board. → not required by law but recommended

When there is a change in PIC, what are the inventory requirements for the *new* PIC?

Complete an inventory of SI-V drugs on the date he or she becomes PIC, and *prior* to opening for business.

How long can a PIC be gone on a scheduled absence before you have to notify the Board and designate a new PIC? Unscheduled absence?

The PIC may be gone 30 days scheduled absence, or 15 days unscheduled absence with no known return date within the next 15 days.

Describe situations in which a drug would be considered adulterated.

- The drug is contaminated (contains "any filthy, putrid, or decomposed substance") or contains a harmful substance in the drug or its container.

- The facilities used to make, pack, or hold it don't conform to cGMP (current good manufacturing practices).

- It doesn't actually contain the strength, quality, or purity of drug that it claims to contain.

- It is mixed or packed with something so as to reduce its quality or strength.

Describe situations in which a drug would be considered misbranded.

- Its labeling is false or misleading.

- The package is missing the name & location of the manufacturer, packer, or distributor.

- The label or labeling is missing required information.

- A habit-forming drug is missing the name and amount of substance, or the statement "Warning – May Be Habit Forming."

The established (generic) name of the drug is not printed on its label prominently and in type at least half as large as that used for the proprietary (brand) name.

- It is missing adequate directions for use or adequate warnings.

- If it is dangerous to health when used in the dosage, or with the frequency or duration prescribed, recommended, or suggested in the labeling or advertising.

- If it is insulin or an antibiotic, and does not have a certificate of release in effect.

- If a trademark, trade name, or other identifying mark was placed on the drug or container with intent to defraud.

Define "proprietary medicine."

A nonprescription, compounded drug promoted to the public under a privately owned trade symbol. It cannot be marketed only to MDs or contain narcotics.

What temperature ranges are considered "freezer temperature," "refrigerator temperature," "cold," "cool," "controlled room temperature," "warm," and "excessive heat"?

- Freezer temperature = -20 to -10C (-4 to 14F)

- Refrigerator temperature = 2-8C (36-46F)

- Cold = <8C (36F)

- Cool = 8-15C (46F)

- Controlled room temperature = 20-25 (68-77F)

- Warm = 30-40C (86-104F)

- Excessive heat = >40C (104F)

Remember: *Fahrenheit Temperature = (1.8 x Celsius Temperature) + 32*

Define "unit dose container."

A unit dose container holds a single dose of a drug not to be administered parenterally.

Define "unit dose package."

A unit dose package contains a particular dose ordered for a patient.

Define "unit dose system."

Multiple drugs in unit dose packaging dispensed in a single container (such as a medication bin) labeled with a patient's name. Directions for administration do not need to be on the labeling.

Define "well-closed container."

A container that protects the contents from extraneous solids during normal shipping and storage.

Define "compliance packaging."

Packaging for solid oral dosage forms designed to assist the user with correct administration or self-administration.

Define "prescription drug."

A drug that is safe to use only under the direction of a practitioner licensed to prescribe or administer.

A patient brings in an unused, unopened bottle of tablets that she picked up earlier today. The patient's doctor started her on a different drug this afternoon. The drug is in the manufacturer's original container, and the seal over the top is intact. Can you re-dispense it to somebody else?

No. It's been out of possession of pharmacy personnel, so you don't have any assurance it was stored properly.

What if the pharmacy delivery driver brought it to her, and she refused it?

If the dispensed drug has not been out of possession of a delivery agent of the pharmacy, it can be returned and re-dispensed.

Can devices be returned for resale?

Yes, in the original sealed packaging.

In a hospital, can a nursing unit return drugs to an on-site hospital pharmacy?

Yes.

Can a nursing home donate unused drugs to a free clinic?

Yes. They may donate the drugs directly to a hospital or pharmacy, under the following conditions:

- The physical transfer is done by a person authorized by the pharmacy.

- The original patient (or his authorized representative) gives written consent, and the nursing home keeps a record.

- The drug name, strength, and expiration date stay on the label, but patient-identifying information is obliterated.

- The transfer includes an inventory list of drugs, strengths, expiration dates, and quantities transferred.

- No outdated drugs are transferred.

- The drugs were not originally dispensed for use by Medicaid, Medicare, and SCHIP patients.

The nursing home returns a prescription to the pharmacy. When the drug is re-dispensed, can it be assigned an expiration date later than the expiration date it had previously?
No.

Who is responsible for determining the suitability of the product for re-dispensing?
The pharmacist-in-charge is ultimately responsible.

If a pharmacy accepts drugs from a nursing home, for the purpose of re-dispensing to the indigent free of charge, what records does the pharmacy need to keep?
- A written agreement with the nursing home

- A current policy & procedure manual describing how medications are delivered and tracked

- A procedure for assuring the integrity of drugs (including proper storage)

- A procedure for assigning a beyond-use date on re-dispensed drugs

How long do pharmacies have to keep records or bulk reconstitution, bulk compounding or prepackaging of drugs?
1 year.   *Bulk cop compounding → 1 year*

What do these records need to contain?
- Drug name & strength, quantity prepared

- Manufacturer or distributor's name and lot/control number

(Drug Name   Name   Date
Strength   lot   expiration
quantity   control #   w/ us)

- Date, expiration date
- Initials of pharmacist verifying the process

If a pharmacy uses automated counting devices in which drugs are removed from manufacturer's original packaging and placed in bulk bins, what record must be kept?

A filling record, which must contain:

- Drug name & strength, quantity prepared
- Manufacturer or distributor's name and lot/control number
- Expiration date
- Any assigned lot number

[handwritten margin note: Drug strength quantity manu lot control # expira. lot # (assign)]

Can you put more than one lot into a bulk bin?

Yes.

If more than one lot is in the bin at the same time, what is the expiration date?

The lot that expires first is used to determine the expiration date.

For bins that can hold more than one lot, what procedure is required to ensure that drugs are in the bin are always in date?

Let the bin "run dry" (all product removed prior to filling) once every 60 days with a record made of the run dry dates.

[handwritten note: Run Dry q 60 days.]

What are acceptable procedures for disposing of unwanted drugs?

Transfer the drugs to another entity authorized to possess or dispose of them.

Destroy by burning in an incinerator.

For SII-V drugs, the PIC must notify the Board in writing 14 days prior to destruction date.

What must the notification to the Board include?

Date, time, manner, and place of destruction, plus the names of the pharmacists who will witness the destruction (the PIC and one pharmacist not employed by the pharmacy.)

What if the destruction date is changed or the destruction doesn't occur?

The Board must be notified.

[handwritten note: ⚹Notification: Date, time, manner, place of destru. and 2 pharm → 1 PIC 1 not employed]

What record must be kept of the destruction?

The DEA drug destruction form is used to record the destruction and kept with the pharmacy's inventory records.

When can pharmacies dispense a drug to a long-term care facility without a valid order?

Never.

A pharmacy is planning to serve a long-term care facility. What is the pharmacy responsible for?

- Ensuring that a valid order is received prior to dispensing any drug
- Training personnel who will be administering the drugs in the dispensing system that the pharmacy provides
- Ensuring that drugs for each patient are kept in the originally received containers and that one patient's medication is not transferred to another patient
- Ensuring that each cabinet, cart or other drug storage area is locked and accessible only to authorized personnel
- Ensuring that the drug storage area is well lighted, big enough to permit storage without crowding, and maintained at appropriate temperature
- Ensuring that poison and drugs for "external use only" are kept in a cabinet separate from other medications
- Providing for the disposition of discontinued drugs
- Ensuring that appropriate drug references are available in the facility units
- Performing a 30-day drug review by a pharmacist for each patient

The doctor discontinues a drug for a patient living in a long-term care facility. What can the facility do with the discontinued drugs?

- Return them to the pharmacy for resale
- Transfer them to another pharmacy authorized to re-dispense them to the indigent
- Destroy them at the pharmacy or at the facility

Destroy @ pharmacy   Destroy @ facility
                            Director
PIC
          Other pharm employee   another employee

30 days per long term

After the medication is discontinued, how long does the pharmacy have to return, transfer, or destroy it?

30 days.

Drugs returned from a long-term care facility are going to be destroyed at the pharmacy. Who must witness the destruction?

The PIC and another pharmacy employee. The drugs may also be transferred to another pharmacy licensed to accept returns for destruction.

Drugs returned from a long-term care facility are going to be destroyed at the long-term care facility. Who needs to witness the destruction?

The Director of Nursing (or the facility administrator, if there is no Director of Nursing) *and* one other person. This could be a pharmacist providing pharmacy services to the facility, or another employee authorized to administer medication.

Destroyed @ facility ⇒ Director of Nursing and One other person.

After the drugs are destroyed, what records are kept?

The original record of destruction is signed & dated by the people witnessing the destruction, and kept at the long-term care facility for 2 years.

A copy is kept at the provider pharmacy for 2 years.

2 years

Some controlled substances are missing from the pharmacy. What is the pharmacy responsible for doing?

- Immediately notify the Board.

- If the exact kind or quantity of drug loss is unknown, do a complete inventory of all SI-V drugs.

- Within 30 days of discovery of the loss, furnish the Board with a listing of the kind, quantity, and strength of the drugs lost.

- Notify the area field division office of the DEA in writing within 1 business day.

- Complete DEA form 106

How long do records of the loss need to be maintained?

2 years.

If controlled substances are lost or stolen while being transported by common contract carrier (such as UPS) between a central fill pharmacy and a retail pharmacy, who is responsible for notifying the DEA?

Whichever pharmacy contracted with the contract carrier.

& who contracted the company

III. Review Prescriptions (30%)

Who can call in a prescription?

The prescriber or his or her authorized agent.

[handwritten: → must have first & last name]

The agent must be specifically directed by the prescriber and must be either:

- An employee under the prescriber's immediate and personal supervision *or*

- An individual holding a valid license allowing administration or dispensing of drugs

Can a written prescription for an SII drug be faxed?

Only for informational purposes. A faxed prescription does NOT count as the original written prescription.

Are there any exceptions to this?

[handwritten: Faxed: Hospice/nursing/home infusion]

Yes, an order for SII drugs for nursing home or home infusion patients may be faxed.

SII prescriptions for hospice patients may be faxed if the prescriber notes on the prescription that the patient is a hospice patient.

These prescriptions must meet same requirements as written prescriptions, including signature.

Can the prescriber decide where to fax the prescription?

It has to be faxed to the pharmacy of the patient's choice.

Where does a faxed prescription need to come from?

- The prescriber's practice location

- A long-term care facility or hospice

Can an oral prescription be faxed?

Yes. It has to have the agent's full name and wording that clearly indicates that the prescription being transmitted is an oral prescription. If the prescription is oral, the prescriber's signature isn't needed.

What other information is required?

The date faxed, the name, address, phone & fax number of the prescriber; and the address, phone number, and fax number of the institution faxed from.

Can a prescriber fax a refill authorization?

Yes.

What information does a faxed refill authorization need to include?

The patient name & address, drug name and strength, quantity, directions for use, prescriber's name, prescriber's signature or agent's name, and date of authorization. *patient name & address (May could be on "back" tag")*

✦MUST be signature

Can a prescriber transmit prescriptions electronically?

Yes.

What information does an electronically transmitted prescription need to include?

The phone number of the prescriber, the full name of the prescriber's agent if other than the prescriber transmitting, and date of transmission.

What are the recordkeeping requirements for electronic prescriptions?

The same as for written prescription records. *2 years from last transaction*

all same requirements

Is a pharmacy allowed to transfer a prescription to another pharmacy?

Yes, if the patient has given permission, and the prescription may be filled or refilled.

Can you transfer a prescription for an SII drug if you haven't filled it?

No transfers on are allowed on SII drugs.

How can a prescription be transferred?

- Orally by direct communication between two pharmacists
- By fax machine
- By electronic transmission *✦Can be by fax!*

What are the responsibilities of the transferring pharmacy?

- Write "VOID" on the face of the invalidated prescription.

[handwritten: DEA of pharmacy must be on transfer w/ controll]

- Record information about the receiving pharmacy (the name, address, and DEA number for controlled drugs).

- For an oral transfer, the name of the pharmacist receiving the prescription.

- Record the date of the transfer.

What are the responsibilities of the receiving pharmacy?

- Write the word "TRANSFER" on the face of the transferred prescription.

- Record required information.

What information is the receiving pharmacy required to record? _[handwritten: original refills need to be on script.]_

- Date of issuance of original prescription

- Original number of refills authorized on the original prescription

- Date of original dispensing, if applicable

- Number of valid refills remaining and date of last dispensing

- Pharmacy name, address, DEA registry number except for Schedule VI prescriptions, and original prescription number from which the prescription information was transferred

- Name of transferring pharmacist, if transferred orally

[handwritten: (electronic not needed)]

How long do the pharmacies need to keep their records?

Both the original and transferred prescription must be maintained for two years from the date of last refill.

[handwritten: 2 years from last refill!]

Can you give a patient a copy of their prescription after you've transferred it?

You may give a prescription marked "For Information Only" to the patient.

Instead or writing the info on the transferred prescription, can the transferring pharmacy record this information electronically?

Yes.

If a patient transfers a prescription from one CVS to another, do both pharmacies need a hard copy on file?

Not if they have a common database that is capable of generating a hard copy of the transferred prescription upon request.

Who can write a prescription for a controlled substance?

A practitioner of medicine, osteopathy, podiatry, dentistry or veterinary medicine who is authorized to prescribe controlled substances, or by a licensed nurse practitioner, a licensed physician assistant, or a TPA-certified optometrist.

When is a prescription not a valid prescription?

- If it is not written for a medical or therapeutic purpose.
- If it is not written for people or animals with whom the practitioner has a bona fide practitioner-patient relationship.

What does the practitioner have to do to ensure a bona fide practitioner-patient relationship?

- Obtain a medical or drug history
- Provide information to the patient about the benefits and risks of the drug being prescribed
- Perform an appropriate examination of the patient, either physically or by using equipment to electronically transmit images and medical records
- Except for medical emergencies, the examination of the patient should be performed by the practitioner, within the group in which he or she practices, or by a consulting practitioner
- Initiate additional interventions and follow-up care, if necessary, especially if a prescribed drug may have serious side effects

Define a bona fide practitioner-patient-pharmacist relationship

The practitioner prescribes, and a pharmacist dispenses, controlled substances in good faith to his patient for a medicinal or therapeutic purpose within the course of his professional practice.

Can a patient get a birth control prescription from a psychiatrist?

Yes. A psychiatrist is an MD, and has the same prescribing authority as any other MD.

Can you accept a controlled substance prescription from an out-of-state practitioner?

Yes, as long as they are authorized to prescribe in Virginia.

If you are authorized to prescribe, does this also mean you are authorized to provide manufacturers' professional samples for controlled substances and devices?

Yes.

What certification does an optometrist need in order to prescribe?

TPA-certification.   TPA

What can an optometrist prescribe? ↓ no SII

- SIII-VI oral analgesics to relieve ocular pain

- Other oral Schedule VI controlled substances to treat diseases of the eye

- Topically applied Schedule VI drugs

- Intramuscular administration of epinephrine to treat emergency cases of anaphylactic shock

If a hospital has a standing order or protocol for administration of influenza or pneumococcal vaccine, does that count as a bona fide practitioner-patient relationship?

Yes.

Give some examples of a practitioner authorizing someone else to possess or administer something under his supervision pursuant to an oral or written order or standing protocol. *Don't worry – it's probably not necessary to memorize all of these.*

- Nuclear medicine technologist administering radiopharmaceuticals

- Registered nurses and licensed practical nurses administering emergency epinephrine or maintaining IV access lines using heparin and sterile normal saline.

- Emergency medical services technicians (EMTs) administering emergency epinephrine.

- Licensed physical therapists to administering topical corticosteroids, topical lidocaine, and any other Schedule VI topical drug.

- Licensed athletic trainers administering topical corticosteroids, topical lidocaine, or other Schedule VI topical drugs, or emergency epinephrine.

- Registered nurses administering tuberculin purified protein derivative (PPD) in the absence of a prescriber.

- A properly trained employee of a school board assisting with the administration of insulin or administering glucagon to a diabetic student, with permission from the parents.

Who can administer vaccines to adults for immunization per Board of Nursing protocol, when a practitioner with prescriptive authority is not physically present?

- Licensed pharmacists

- Registered nurses

- Licensed practical nurses under the immediate and direct supervision of a registered nurse

What can a dentist authorize to be administered by either a dental hygienist or by an authorized agent of the dentist?

- Schedule VI topical drugs

- Oral fluorides

- Topical oral anesthetics

- Topical and directly applied antimicrobial agents for treatment of periodontal pocket lesions

- Any other Schedule VI topical drug approved by the Board of Dentistry

- Schedule VI nitrous oxide and oxygen inhalation analgesia

- Schedule VI local anesthesia, to patients 18 years of age or older

Can a prescription be written by a medical intern or resident who is works only in a hospital?

Yes, but only within their duties as part of the residency program.

It needs the prescriber's signature, the legibly printed name, address, and telephone number of the prescriber and an identification number assigned by the hospital.

DEA of the hospital w/ suffix of the particular prescriber

Does a medical intern or resident have a DEA number?

They have an identification number – the DEA number of the hospital pharmacy, plus a suffix assigned by the institution.

If you decline to fill a prescription for any reason other than the unavailability of the drug, what do you have to do?

On the back of the prescription, write the following   *Back of Rx*

- "Declined"
- The name, address, and telephone number of the pharmacy
- The date the prescription was declined
- The pharmacist's signature

What factors does the US Attorney General consider when adding a drug to or removing it from Schedules I-V?

- Its actual or relative potential for abuse
- Scientific evidence of its pharmacological effect, if known
- The state of current scientific knowledge regarding the drug
- Its history and current pattern of abuse
- The scope, duration, and significance of abuse
- What, if any, risk there is to the public health
- Its potential to produce psychic or physiological dependence
- Whether the substance is an immediate precursor of a drug that is already scheduled

Can the Virginia Board of Pharmacy schedule or deschedule drugs?

Yes, as long as the drug is not OTC or already deemed a controlled substance by the FDA.

What factors does Virginia consider when scheduling drugs?

The same ones the US Attorney General considers.

When are you prevented from substituting a generic drug for the brand name drug in filling a prescription?

- If the generic and brand are not therapeutically equivalent
- If the prescriber wrote "brand medically necessary" on the prescription

↳ MUST write on Rx
cannot be a check box.

- If the prescriber orally requested the brand name drug

- If the patient requested the brand name drug

- If the generic is more expensive than the brand

Can a prescriber disallow substitution by checking a "Dispense as Written" box on the prescription?

No. This was allowed prior to July 1, 2006, but is not currently allowed.

What do you have to do if you dispense a generic drug instead of the brand name drug prescribed?

- Inform the person who purchases the drug.

- On the patient record and prescription label, indicate the brand name (or, if a therapeutically equivalent drug product, the name of the manufacturer or distributor).

- Label the drug with the product dispensed followed by "generic for" and the brand name.

& MUST HaVE BoTH On the label

If a doctor has a DEA number, does this mean that he or she can prescribe controlled substances to treat patients addicted to narcotics?

No. Any doctor may use narcotics for pain control, but treating addiction requires special registration.

What registration is required for a doctor to use SII drugs to treat narcotic addiction?

To use SII drugs (e.g., methadone), the doctor must obtain separate registration from DEA as a narcotic treatment program and have a separate DEA number for this purpose. *Registration Separation*

What restrictions apply to a doctor who has obtained separate DEA registration to treat narcotic addiction?

- The total number of patients can't exceed 30

- The doctor must be able to refer patients for appropriate ancillary services

- The drugs used have to be approved for "maintenance treatment" or "detoxification treatment"

not over 30 pts.

What registration is required for a doctor to use SIII-V drugs to treat narcotic addiction?

There is one SIII drug approved for treatment of narcotic addiction, buprenorphine (Subutex, Suboxone). The doctor needs to receive a waiver under the Drug Addiction Treatment Act of 2000 (DATA 2000). The Substance Abuse and Mental Health Services Administration (SAMHSA) maintains a list of doctors authorized to prescribe buprenorphine. They may prescribe using a DEA number that is the same as their usual DEA number, but with an X in place of the first letter.

What about in an emergency? If a patient is addicted to narcotics and cannot be admitted to a treatment program until the next day, is there anything the family physician can do to ease withdrawal symptoms?

Yes. The "three day rule" allows the doctor to administer (but not prescribe) narcotics for the purpose of relieving acute withdrawal symptoms while arranging for the patient's referral for treatment, under the following conditions:

- Only one day's medication may be administered or given to a patient at a time.

- The treatment cannot last longer than 72 hours.

- The 72-hour period cannot be renewed or extended.

Administer only 72 hours [handwritten]

*Relieving acute withdrawal [handwritten]

Can methadone be administered by a narcotic treatment program? Dispensed? Prescribed?   *not prescribed [handwritten]*

Methadone can be administered and dispensed, but not prescribed. A pharmacist can fill a prescription for methadone if it is used for pain control, but not for treatment of addiction.

Can buprenorphine be administered by a physician? Dispensed? Prescribed? *Subox. Can be prescribed [handwritten]*

Buprenorphine can be administered, dispensed, and prescribed.

Can a physician or authorized hospital staff administer or dispense narcotic drugs in the hospital for maintenance or detoxification?

Yes.

What is required for a prescription for gamma-hydroxybutyric acid?

The medical need written on the prescription.

* GHB → CIII [handwritten]

*Medical Need [handwritten]

Can a prescription be written in pencil?

No – it may be written in ink, individually typed, or printed only.

What information about the prescriber does a prescription need to contain?

Name, address, phone number, DEA number.

Do you need a DEA number for all prescriptions?

Not if the prescription is for a SVI drug.

Not for non-scheduled → NPI required?

What information about the patient does the prescription need to contain?

First and last name and address.

Does the prescriber need to write the address on the prescription?

No, it can be written in by the dispenser or kept in the patient's electronic record in the pharmacy.

Can you accept a postdated SII prescription?

No. It must be dated and signed on the day written only

Can you accept a postdated SIII-VI prescription?

No, it must be dated on the day written only.

Can a prescriber use a preprinted prescription for SII drugs?
No.

Can a prescriber use a preprinted prescription for SIII-V drugs?
No.

Can a prescriber use a preprinted prescription for SVI drugs?

Yes, this is the only schedule where preprinted prescription forms are allowed.

Can a prescriber put more than one prescription on a sheet?

Generally, no.   *On long term & hospital*

The only patients who can receive multiple prescriptions on a sheet are:

- Patients in hospitals or long-term care facilities

Only preprinted on schedule VI

- Patients receiving home infusion services

- Hospice patients

- Patients whose prescriptions will be filled at pharmacies operated by one of these departments: Corrections, Juvenile Justice, Health, Mental Health, or Mental Retardation and Substance Abuse Services

- Patients residing in detention centers, jails, or work release centers

Can faxed prescriptions ever be considered *original written* SII prescriptions?

Yes, but only for two types of patients:

- Residents of long-term care facilities

- Patients receiving infusion therapy from a home infusion pharmacy

A veterinarian has a DEA number. Does that mean he can prescribe you oxycontin?

No, a veterinarian can't prescribe for humans.

When does a pharmacist conduct a prospective drug review?

- Before each new prescription is dispensed or delivered.

- Before refilling a prescription, if needed based on the pharmacist's professional judgment.

What problems are screened for in a prospective drug review?

- Therapeutic duplication

- Drug-disease contraindications

- Drug-drug interactions, including nonprescription drugs

- Incorrect drug dosage

- Incorrect duration of drug treatment

- Drug-allergy interactions

- Clinical abuse or misuse

What kind of review of drug therapy is required for patients who spend a long time in the hospital?

Drug therapy must be reviewed monthly. At a minimum, this includes reviewing any irregularities in drug therapy, drug interactions, drug administration, or transcription errors.

What kind of review of drug therapy is required in a long-term care facility?

A monthly review, the same as in a hospital.

It is legal to outsource prescription processing to another pharmacy?

Yes. This is called a central fill pharmacy arrangement. The pharmacy that does the prescription processing is called the remote processor.

In a central fill pharmacy, what is the remote processor responsible for?

The remote pharmacy may do everything except dispense.

Activities can include:

- Receiving, interpreting, analyzing, or clarifying prescriptions
- Entering prescription and patient data into a data processing system
- Transferring prescription information
- Performing a prospective drug review
- Obtaining refill or substitution authorizations, or otherwise communicating with the prescriber concerning a patient's prescription
- Interpreting clinical data for prior authorization for dispensing
- Performing therapeutic interventions
- Providing drug information or counseling concerning a patient's prescription to the patient or patient's agent

It is legal to outsource prescription processing to an out of state pharmacy?

Yes.

What are the requirements for outsourcing prescription processing to a remote location?

- The two pharmacies must have the same owner or there must be a written contract between the pharmacies.

★Same owner → Common electronic file
policy/procedure manual

- The pharmacies must share a common electronic file.

- Both locations must have a policy and procedure manual.

- Records of processing tasks and the people performing tasks – I call this the "who did what where" record.

What must be contained in the "who did what where" record?

- Each processing task

- The identity of the person performing each task. This includes both pharmacists and technicians.

- Location where each task was performed

What are the requirements for maintaining the "who did what where" record?

- The record may be maintained separately by each pharmacy or in a common electronic file.

- The primary dispensing pharmacy must be able to retrieve at least the last two years worth of records.

Record maintained separately

What are the requirements for outsourcing prescription processing to an out of state pharmacy?

- Both pharmacies must follow VA law for supervision of technicians and activities restricted to pharmacists.

- Technicians must be certified in VA or possess "substantially equivalent" credentials.

- A pharmacist licensed in Virginia must perform a final check for accuracy. This pharmacist can be at the remote pharmacy or the dispensing pharmacy.

Pharmacist licensed in VA → final check

Is a pharmacy required to notify patients if outsourcing prescription processing?

Yes.   *Must notify patient.*

How?

- A one-time written notification

- A sign posted in the pharmacy in a location that is readily visible to the public

must have written notification / sign posted.

What does the notice have to include?

- The name of the pharmacy providing central or remote prescription processing

- If the pharmacy uses a network of pharmacies under common ownership, this must be disclosed

What must be spelled out in the policy and procedure manual?

- The responsibilities of each pharmacy

- A list of the name, address, telephone numbers, and permit/registration numbers of all pharmacies involved in central or remote processing

- Procedures for protecting the confidentiality and integrity of patient information

- Procedures for ensuring that pharmacists performing prospective drug reviews have access to appropriate drug information resources

- Procedures for maintaining required records

- Procedures for complying with all applicable laws and regulations to include counseling

- Procedures for objectively and systematically monitoring and evaluating the quality of the program to resolve problems and improve services

- Procedures for annually reviewing the written policies and procedures for needed modifications and documenting such review

If a pharmacy is performing remote prescription processing for a hospital or long-term care facility, one additional procedure must be spelled out in the manual. What is it?

Procedure for authorizing the administration of the drug by the appropriate staff member.

Can a pharmacist licensed in VA access their employer pharmacy's database from a remote location to process prescriptions?

Yes, as long as the pharmacy establishes controls to protect the privacy and security of confidential records.

If a hospital or long-term care facility outsources prescription processing, are the rules the same as for a central fill pharmacy?

There is one exception. Any pharmacist participating in remote prescription processing must be licensed in VA.

Does a hospital or long-term care facility that outsources need a "who did what where" record?

Yes.

It has to be available by prescription order or by patient name.

It has to be available for 2 years.

IV. Dispensing and Distribution (25%)

Define "repackaged drug."

A drug removed from the manufacturer's original package and placed in different packaging.

Define "safety closure container"

A container that meets the requirements of the Poison Prevention Packaging Act of 1970:

- Out of a group of 200 children age 41-52 months, 85% can't open the package in 5 minutes
- After a demonstration of how to open the package, and 5 more minutes, 80% of the children still can't open it
- 90% of adults *can* open and close it

Define "special packaging."

Packaging that is designed to be difficult for children under 5 years of age to open.

Define "compounding."

Combining two or more ingredients by a pharmacist in these cases:

- Pursuant to a prescription
- In expectation of receiving a valid prescription based on observed prescribing practices
- For a practitioner to dispense in the course of professional practice
- For research, teaching, or chemical analysis

A "drug" or "device" is something intended for what?

Diagnosis, cure, mitigation, treatment, or prevention of disease in humans or animals, or to affect the structure or any function of the body.

A "drug" is also an article recognized in which publications?

- The official United States Pharmacopoeia National Formulary
- The official Homeopathic Pharmacopoeia of the United States

- A supplement to one of these pharmacopoeia

Define "dispense."

Delivering a drug to an ultimate user, research facility, or practitioner pursuant to a lawful order of a practitioner. If a doctor gives medications to patients to take with them from the office, that's dispensing. Transporting drugs from the doctor's Arlington office to the Falls Church office is not dispensing.

Define "administer."

Direct application of a controlled substance to the body of a patient or research subject by a practitioner or his or her agent, or by the patient under the direction of the practitioner.

Define "label."

Written, printed or graphic matter on a drug's immediate container. If something is required to be on the label, it must be visible on or through the outside container.

Define "labeling."

The drug label, plus materials accompanying the drug.

What are the requirements for an SII prescription? *Name & address of the patient.*

- Written, not oral

- Signed and dated on the day written – postdated prescriptions are not allowed

- The name, address of the patient must be on the prescription

- The name, address, and DEA number of practitioner must be on the prescription *Signature*

Can you accept a SII prescription for an animal?

Yes. The same rules apply as for a human SII prescription.

What information must be contained on an SII prescription for an animal?

The name & address of the owner, and *species* of the animal.

name + address of the species.

---

need the actual copy in 7 days

Can you ever accept an oral prescription for an SII drug in a community pharmacy?

Yes, but only in emergencies. The prescriber must provide the pharmacy with the original signed written prescription within (7) days after authorizing the emergency prescription.

The pharmacist attaches the signed prescription to the oral emergency prescription. If the prescriber doesn't provide the signed prescription, the pharmacist must notify the DEA.

in a hospital → must be signed in 72 hours

Can you accept an oral order for an SII drug in a hospital?

In the hospital, a pharmacist or nurse can take an oral order for an SII drug, but it must be signed by the prescriber within (72) hours.

 What needs to be on the label of an SII prescription?

- Prescription serial number or name of the drug

- Date of initial filling

- Pharmacy or pharmacist name & address

- Patient name & address (or owner info and species, if an animal prescription) 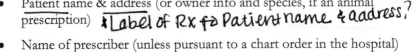 *Label of Rx → Patient name & address?*

- Name of prescriber (unless pursuant to a chart order in the hospital)

- Directions for use

What are the requirements for SIII-VI prescriptions?

- Written *or* oral – a written prescription requires a signature, oral doesn't

- Signed and dated on the day written – postdated prescriptions are not allowed

- Name & address of the patient

- Name, address of owner & species of animal if an animal prescription

- Name & address, of practitioner

- The prescriber's DEA number is *not* required

† not required per VA law.

The patient needs a refill for a SVI drug, and the prescriber isn't available. What can you do?

An SVI drug may be refilled without authorization from the prescriber if the pharmacist has a made reasonable effort to contact him or her, and the patient's health would be in imminent danger without the drug.

† w/o authorization

The pharmacist must inform the patient that the refill is being made without authorization, and inform the prescriber of the refill.

On the back of the prescription, the pharmacist must note the date and quantity of the refill, the prescriber's unavailability and the rationale for the refill.

↳ on the back of the Rx - quanity given

Do you really need to have the prescriber address on an SVI prescription?

If the information is readily retrievable within the pharmacy, it doesn't need to be on an SVI prescription. ~~address~~ *not a nec.*

Can a nurse call in a prescription for the doctor?

Yes, for SIII-SVI drugs. The pharmacist should write down the full name of the nurse transmitting the prescription.

Can you get a refill on an SII drug?

No refills on SII drugs!

How many refills are allowed on SIII-IV drugs?

5 refills in 6 months.

What about partial fills? *Partial not exceeding total*

Partial refills are OK – this means that if the prescription is for clonazepam 0.5 mg po daily #30 with 5 refills, the patient can get 15 tabs at a time every 2 weeks for the whole 6 months. The patient can't get any more after the 6 months are up, even if they did not use all of the refills.

When can you do a partial fill of an SII drug?

If the pharmacy can't supply the full quantity, a partial fill is allowed. The rest must be dispensed within 72 hours. If it can't be, the pharmacist must notify the physician. No more can be dispensed after the 72 hours without a new prescription.

Can a pharmacy dispense partial fills to patients in a long-term care facility?

Yes, for up to 60 days.

The long-term care facility must keep a record of the date of the partial dispensing, quantity dispensed, remaining quantity authorized to be dispensed, and the identification of the dispensing pharmacist.

60 days only in long term/hospice

Can a long-term care facility maintain computerized records of current SII prescriptions?

Yes, if the system allows immediate (real-time) updates every time a partial is dispensed, and can produce output showing the original prescription number, date of issue, identification of prescribing practitioner, identification of patient, identification of the long-term care facility, identification of drug authorized, and a listing of partial dispensing under each prescription.

Can a terminally ill patient get partial fills of an SII drug?

Yes, for up to 60 days.

The practitioner must classify the patient as terminally ill, and the pharmacist verifies and records this on the prescription. Prior to each subsequent partial fill, the pharmacist must determine that it is necessary.

The pharmacy must maintain a record of the date, quantity dispensed, remaining quantity authorized to be dispensed, and the identity of the dispensing pharmacist.

Record, quantity, remaining/dispensed and the Identity of the pharmacist

How long is a prescription for a SVI drug valid?

It's valid for one year after the date of issue. This means that you can't give refills for more than one year after the date of issue unless the prescriber specifically authorizes dispensing or refilling for a longer period (up to 2 years).

Can be authorized for 2

What information does the pharmacist have to record when dispensing a prescription?

The date of dispensing, and the dispensing pharmacist's initials. Electronic records are acceptable.

Is the pharmacy required to keep the hard copies of all prescriptions? What if the prescriptions are oral?

Yes, a hard copy of everything must be kept (including oral prescriptions which have been reduced to writing). The records are filed chronologically and kept for 2 years.

Can you keep all the prescriptions together?

No, SII prescriptions must be stored separately.

Can the pharmacy store SIII-V prescriptions with the SVI ones?

Yes, but the SIII-V prescriptions need to be "readily retrievable." There are two ways to accomplish this:

- The SIII-V prescriptions have a red letter C at least 1 inch high stamped in the lower right corner, OR

- The pharmacy uses an electronic data processing system that allows identification by prescription number and retrieval of original documents by prescriber's name, patient's name, drug dispensed, and date filled

Does an order on a chart need to contain all the same information as a written prescription?

No, under the following conditions:

- The information is contained in other readily retrievable records of the pharmacy

- The pharmacy's policy and procedure manual that sets out where this information is maintained and how to retrieve it

- The minimum requirements for chart orders consistent with state and federal law and accepted standard of care

How are chart orders filed?

Chronologically.

Charts orders may be filed using another method if:

- Dispensing data can be produced showing a complete audit trail for any requested drug for a specified time period

- Chart orders are readily retrievable upon request

- The filing method is clearly documented in a current policy and procedure manual

How is a chart order filed if it contains both an order for a SII drug and orders for drugs in other schedules?

If the drug is floor stock, no additional filing is necessary

If the drug is dispensed from the pharmacy, the original order is filed with records of SII drugs, and a copy of the order is placed in the file for other schedules.

Is the pharmacy required to keep hard copies of prescriptions?

Yes. For SVI prescriptions, an electronic image is sufficient if it's legible and can be made available within 48 hours.

The pharmacy also doesn't need a hard copy if the prescription is an electronic automated transmission from the prescriber – in that case, the automated transmission is the hard copy.

How long must a pharmacy computer system be able to store information?

The system needs to maintain two years worth of information.

at least 2 years

A pharmacy keeps its prescription records in an automated data processing system. Is the pharmacy required to print anything out?

Yes, the pharmacy must make a daily printout summarizing all transactions that occurred during the day. The printout is signed by each pharmacist who dispensed prescriptions that day.

A log book may be used in lieu of a printout.

Must have daily printout of all transactions

What kind of packaging can a drug be dispensed in?

- USP-NF approved packaging

- Well-closed container

- Compliance packaging when requested by the patient, or for use in hospitals or long-term care facilities

If a patient requests non-special packaging (non-child-resistant), does the pharmacy have to document this?

Yes. A record of the request has to be kept for two years.

What does the Code of Virginia require to appear on the prescription label?

- Drug name, strength

- Brand and generic names (unless administered in a hospital or long-term care facility by a licensed person)

- Number of dosage units or, if liquid, the number of milliliters dispensed

+ Brand & generic names must be on label

What does the Code of Federal Regulations require to appear on a prescription label?

For SII-IV drugs, "Caution: Federal law prohibits the transfer of this drug to any person other than the patient for whom it was prescribed."

This statement is not required to appear on the label of a controlled substance dispensed for use in blinded clinical trials.

+ only exception is in trials

Before dispensing a new prescription, what does a pharmacist do?

A prospective drug review. For refills, prospective drug reviews only need to be done when the pharmacist deems necessary, based on clinical judgment.

When dispensing a new prescription, the pharmacist has to offer what?

The pharmacist has to offer to counsel. For refills, the offer to counsel only needs to be made when the pharmacist deems necessary, based on clinical judgment.

How can the offer to counsel be made?

- Face-to-face
- A sign
- A notation on the bag
- By phone

What info does the pharmacy need to make "reasonable efforts" to collect about their patients?

- Name, address, date of birth, gender, and phone number
- Medical history including allergies and drug reactions
- Current medications and devices
- Failure to accept pharmacists offer to counsel

If you don't document refusal of the offer to counsel, what's the implication?

That the offer was accepted and counseling was provided.

Is a pharmacist required to counsel patients in a hospital or nursing home?

No. Not required in the hospital/nursing home.

When a practitioner (or person supervised by them) mixes, dilutes, or reconstitutes a drug for administration to a patient, is this considered compounding?

No.

Is the pharmacy required to put a beyond-use date on compounded products?

Yes, in accordance with USP-NF.

MUST put beyond use date.

Can a pharmacy compound without a prescription in anticipation of receipt of prescriptions based on a routine, regularly observed prescribing pattern?

Yes.

What information needs to appear on a compounded product?

- Name and strength of the medication or a list of the active ingredients and strengths   *↑ name of all active*
- Quantity   *↑ quanity*
- The pharmacy's assigned control number that corresponds with the compounding record
- An appropriate beyond-use date

Can a pharmacy distribute compounded drugs for resale?

No.

If a pharmacy delivers compounded drugs to a hospital or nursing home, is this considered distributing the drugs for resale?

No. If the drugs are compounded pursuant to valid prescriptions, the pharmacy can deliver them to a nursing home or hospital.

Can the pharmacy provide compounded drugs to a practitioner?

Yes, to administer to their patients in the course of their professional practice, either personally or under their direct and immediate supervision.

Compounded allowed to practitioner

What must appear on the label when a pharmacy provides a compounded drug to a practitioner?

- "For Administering in Prescriber Practice Location Only"
- The name and strength of the compounded medication or list of the active ingredients and strengths
- Quantity
- The facility's control number
- An appropriate beyond-use date

Do pharmacists have to do the compounding, or can a technician do it?

A technician may compound if a pharmacist supervises the process. The pharmacist must do a <u>final</u> check for:

- Accuracy
- Correct ingredients
- Correct calculations
- Accurate and precise measurements
- Appropriate conditions and procedures
- Appearance of the final product

Can a pharmacy compound with ingredients that are not considered drug products?

Yes, in accordance with USP-NF standards.

If a drug has been withdrawn from the market by the FDA for safety reasons, can you compound with it?

Not for human use.

Can you compound a product that is essentially a copy of a commercially available drug product?

Not regularly. However, you can do it if:

- There is a <u>change in</u> the product <u>ordered</u> by the prescriber for an individual patient
- The <u>product is unavailable</u> from the manufacturer or supplier
- The pharmacist is <u>mixing two or more commercially available products</u> regardless of whether the end product is a commercially available product

What records do you need to keep of a product compounded for an individual patient?

- Name and quantity of all components
- Date of compounding and dispensing
- Prescription number or other identifier of the prescription order
- Total quantity of the finished product

Name, quanity, date (compounding + dispensing), Rx #
the total of the finished product, initial of compounder & RPh.

- Signature or initials of the pharmacist or pharmacy technician performing the compounding

- Signature or initials of the pharmacist responsible for supervising and checking

When compounding in anticipation of prescriptions, what must be recorded?

galen

- Generic name of each component

- Manufacturer or brand name of each component

- The manufacturer's lot number and expiration date for each component (if unknown, the source of the component)

- The assigned lot number, if subdivided

- The unit or package size and the number of units or packages prepared

- Beyond-use date

The criteria for establishing the beyond-use date shall be available for inspection by the Board.

✦ the established data m records.

Does the pharmacy need to maintain a complete compounding formula listing all procedures, necessary equipment, necessary environmental considerations, and other factors in detail?

Yes, if the instructions are necessary to replicate a compounded product or where the compounding is difficult or complex and must be done by a certain process in order to ensure the integrity of the finished product.

✦ all info needed

Is any kind of quality assurance required for compounding?

Yes, the pharmacy must have a formal written quality assurance plan that describes how compounding is monitored and evaluated in compliance with USP-NF standards. This must include training of personnel involved, and initial and periodic competence assessment.

training of personnel & competence assessment.

If a hospital or long-term care facility uses a unit dose dispensing system, can they store medication outside of the pharmacy?

Yes, as long as the drugs are locked up.

✦ MUST BE LOCKed!

How is a hospital or long-term care facility required to store drugs for an individual patient?

In an individual drug drawer or tray, labeled with the patient's name and location. All unit dose drugs intended for internal use shall be maintained in the

✦ all individual

patient's individual drawer or tray unless special storage conditions are necessary.

The nurse in a hospital would like to have a "back-up dose" of a drug available for a specific patient. Is this allowed?

Yes. Only one back-up dose unit may be maintained in the patient's drug drawer along with the other drugs for that patient.

May have one back updace

Is it ever possible for unlicensed people to administer medications?

Yes, in a long-term care facility this might happen.

Can a prescriber verbally order an SII drug in the hospital?

Yes, but a nurse or pharmacist must immediately reduce it to writing and get the order signed within 72 hours.

72 hours

Can a pharmacy technician transcribe the prescriber's drug orders to a patient profile card, fill the medication carts, and perform other duties related to a unit dose distribution system?

Yes, provided these are done under the personal supervision of a pharmacist.

Doses in the unit dose system need to be labeled with what?

Drug name, strength, lot number and expiration date when indicated.

Drug, strength, lot # & expiration

When a drug cart is filled in a hospital, what records does the pharmacy need to maintain?

Date of filling, location of cart, initials of person who filled it, initials of the checking pharmacist.

How long is the pharmacy required to keep these records? — *Drug cart?*

1 year.

If a pharmacy uses a unit dose dispensing system they must keep a dispensing record. What kinds of records are acceptable for drugs dispensed to a specific patient's drug drawer?

- A patient profile or medication card – the record of dispensing is entered at the time the drug drawer is filled.

- An electronic "fill list" that lists drugs dispensed to patients' drug drawers.

Records in the drug cart fills - 1 year
Schedule II-V: 2 years

How long do these records need to be maintained?

For SII-V drugs, for (two) years.

What about drugs that the pharmacy distributes as floor stock within an institution? What dispensing record is required?

Records of administration are acceptable for drugs distributed as floor stock.

When providing unit dose systems to hospitals or long term care facilities, how many days worth of drugs can be dispensed at a time?

- If only *licensed* persons administer drugs, (a 7-day) supply in solid oral dosage form.

- If *unlicensed* persons administer drugs, (a 72 hour) supply in solid oral dosage form.

If unlicensed persons will be administering drugs within the long-term care facility, what else is the pharmacy responsible for providing?

- Training on the unit dose system

- A medication administration record, listing each drug to be administered with full dosage direction – no abbreviations

- Drugs placed in slots in a drawer labeled or coded to indicate time of administration

Name four situations where a pharmacy may provide drugs that leave the pharmacy but *aren't* for a specific patient:

- Floor stock (e.g., drugs in an Accudose or Pyxis machine in a hospital)

- Drug kit for licensed Emergency Medical Services program

- Emergency drug kit (e.g., a crash cart in a hospital or other facility where licensed practitioners administer drugs)

- Stat drug box (e.g., in a long term care facility, for situations where waiting for the first dose might endanger the patient)

What schedule drugs can the pharmacy supply as floor stock?

Any schedule – a receipt is needed for SII-V.

all schedules can be placed in floor stock

What records does the pharmacy need to maintain of drugs supplied as floor stock?

No records are required for SVI.

For SII-V, the pharmacy needs:

Dispensing pharmacist & receiving nurse.

- A delivery receipt – including the hospital unit receiving drug, and signatures of the dispensing pharmacist and receiving nurse

- A record of disposition/administration – it must be returned to the pharmacy within *three months* of the drug's use.

the delivery / admin.

So can the pharmacy just file these records and forget about them?

No! There are significant auditing requirements. Periodically (e.g., monthly) the PIC (or someone the PIC assigns) has to:

- Match returned records with dispensing receipts – to make sure everything is accounted for. *& monthly make sure delivery & rec. match*

- Audit for completeness – records must include patient name, dose, date/time of administration, signature/initials of person administering drug, and the date the record was returned. *patient, dose, time/date signature & date receipt return*

- Verify the inventory is correct – that additions and deductions from inventory are correctly calculated, and sums are correctly carried from one record to the next.

- Verify that doses documented on administration records are reflected in the medical record. *reflected in the medical record properly.*

Whew! You've finished the auditing – now what? How should the records be filed? How long should they be kept? May they be kept offsite?

Initial the records and file them chronologically. They must be kept 2 years. They may be stored offsite if they can be retrieved within 48 hours.

2 years → retrieved in 48 hours.

The nurse returns a record of administration that shows both morphine and Ambien administration to the same patient. Is this acceptable? Doesn't Virginia require keeping SII records separate from the other schedules?

As long as the SII drugs are listed in a separate *section*, they may be listed on the same page as other schedules of drugs. *& listed in separate.*

Can the pharmacy provide a drug kit for a licensed emergency medical services (EMS) agency?

Yes, but it has to be sealed to preclude any possibility of loss of drugs.

Who orders and administers the drugs from one of these drug kits?

- An authorized practitioner orders them (either orally or with a written standing order).

- An emergency medical technician (EMT) administers them.

- The practitioner signs the orders afterwards.

- If the practitioner refuses, the operational medical director for the EMS agency signs it and returns the orders with the kit to the pharmacy within 7 days.

the practitioner needs to sign (handwritten)

What should the EMT do with the kit after it has been opened?

Return it to the pharmacy for exchange.

Does the pharmacy have to keep a record of the kits provided and drugs administered?

Yes, the pharmacy keeps a record of the drugs administered. It's kept for 2 years, just like any other prescription or order.

A record of the kit exchange is kept for a year.

the record of exchange is 1 year (handwritten)

Can the pharmacy prepare an emergency drug kit for a facility (e.g., a hospital crash cart, or an emergency kit for a long-term care facility)?

Yes, under the following conditions:

- Only licensed persons administer drugs at the facility

- Only drugs necessary for patient survival are kept in the kit

- Only drugs for injection or inhalation, or sublingual nitroglycerin, are kept in the kit

Who decides what is in an emergency drug kit for a facility?

The pharmacist, in consultation with the medical & nursing staff of the facility.

What are the requirements for sealing an emergency drug kit?

It must be sealed to preclude possible loss of the drug. Once the seal is broken, it must not be able to be resealed without detection. There are two kinds of acceptable seals:

- Seals with a unique numeric or alphanumeric identifier. The pharmacy keeps records of seals currently used on kits it has provided.

- A built in mechanism preventing resealing or relocking once opened, except for by the pharmacy.

only able to be resealed by pharmacy

A nurse opens the kit. What does the nurse have to do afterwards?

- Fill out the form inside the kit with the name of person opening the kit, the date, time, name & quantity of items used.
- Make sure any drug used from the kit is covered by a prescription, signed by the prescriber when legally required, within 72 hours.

72 hours to get approved.

Can a pharmacy provide a stat-drug box to a long-term care facility?

Yes, but drugs at the facility may only be administered by persons licensed to administer.

What kinds of drugs are permitted in a stat-drug box?

If delaying therapy may result in harm to the patient, then the drug may be kept in a stat-drug box.

No SII drugs are allowed.

no SII only SIII-V One in each class

No more than one SIII-V drug in each therapeutic class is allowed.

No more than 5 doses of each SIII-V drug are allowed.

Who decides what is in the box?

The same people who decide what is in an emergency drug kit – the pharmacist in consultation with the medical & nursing staff of the facility.

What are the sealing requirements?

The requirements are the same as for an emergency drug kit.

A nurse opens the stat drug box. What does the nurse have to do next?

- Fill out the form inside the kit with the name of person opening the kit, the date, time, name & quantity of items used.
- Make sure any drug used from the kit is covered by a prescription, signed by the prescriber when legally required, within 72 hours.
- Return the opened box to the pharmacy.

What additional records need to be kept for stat drug boxes?

- A list of the contents of the box – attached to the box, and included in the facility's policy & procedure manual

Records of all manded the policy & procedure & expiration date.

- An expiration date – the earliest date that any drug in the box will expire

Can a pharmacy provide an emergency drug box to a long-term care facility?

Yes, if only people licensed to administer are administering drugs.

Can a pharmacy provide a stat drug box to a long-term care facility?

Yes, if only people licensed to administer are administering drugs.

Can a pharmacy provide floor stock to a long term care facility?

Yes, if only people licensed to administer are administering drugs.

What drugs can be provided to a long-term care facility as floor stock?

- Intravenous fluids
- Irrigation fluids
- Heparin flush kits
- Medicinal gases
- Sterile water
- Saline
- Prescription devices.

IV
irrigation
Heparin
Meaicnal
Sterile
Saline
RX devices

Who decides what the long term care facility may have as floor stock?

The pharmacist, in consultation with the medical & nursing staff of the facility.

Can a pharmacy provide an emergency box or a stat-drug box to a correctional facility?

Yes, if the facility employs one or more full-time physicians, registered nurses, licensed practical nurses, physician assistants or correctional health assistants.

the facility → one / more full time's

Hospitals may use automated dispensing systems under certain conditions. What are the conditions?

- Drugs must be under control of a pharmacy.
- The PIC must have established procedures for stocking, storage, security and accountability.
- Drugs are only removed pursuant to a valid order or prescription.

under control of PIC → must have policy / procedure. Only w/ valid order.

- Drugs are in the original sealed packaging or unit-dose containers packaged by the pharmacy.

- The machine must be able to produce a hard-copy record of drugs distributed, including the identity of the patient and of the nurse withdrawing the drug.

- Personnel are only allowed to access the dispensing machine using a personal access code.

- Proper use of the dispensing machines is set forth in the pharmacy's policy and procedure manual.

Who is ultimately accountable for drugs dispensed from an automated dispensing system?

The PIC, but the actual filling and stocking can be delegated to another pharmacist or pharmacy technician as long as they are an employee of the facility, and are properly trained. If a pharmacist does it, the pharmacist is responsible. If a technician does it, the PIC takes responsibility.

What are the monitoring requirements for automated drug dispensing systems?

automated
monthly

- They must be audited monthly.

- Operation and maintenance must be periodically reviewed.

- They must be periodically inspected.

A technician is about to remove drugs from the pharmacy to load the automated dispensing machine. What record must be made?

A delivery record, including:

- Date

- Drug name, dosage form, strength and quantity

- Hospital unit

- A unique identifier for the device receiving the drug

- Initials of the person loading the automated dispensing machine

- Initials of the pharmacist reviewing the transaction.

Delivery: Date, drug name, dosage form, strength & quantity, hospital unit, unique identifier for device, initials of loading, initials of pharmacist reviewing

What is the technician responsible for doing when loading the automatic dispensing machine?

- Obtain a signature from a nurse (or other person authorized to administer drugs) for all SII-V drugs.

- Verify that the count of that drug in the automated dispensing device is correct.

- If there are any discrepancies, note them on the dispensing record and report the situation to the pharmacist in charge.

note if any discrepancies

Can the nurse receiving SII-V drugs sign electronically?

Yes, under the following conditions:

- The electronic signature is a unique identifier and restricted to the individual receiving the drugs.

- The record of unique identifiers is maintained in a read-only format which cannot be altered after the information is recorded.

- The electronic record is readily retrievable and maintained for two years.

- The system can produce a hard-copy printout of the record upon request.

electronic → 2 years

Who performs auditing of the automated dispensing machines, and how often?

The PIC or someone the PIC designates audits the machines monthly.

The audit is initialed and dated by the pharmacist who conducted the review – or by a pharmacist if a technician conducted the review.

What is included in the audit?

- The count of SII-V drugs dispensed from pharmacy is checked against the quantity loaded into the machines.

- The count on hand is checked.

- A 24-hour sample of administration records from each device checked for possible diversion by fraudulent charting.

- Medical records are checked to ensure a valid order exists for a random sample of doses recorded as administered.

- Records are checked for compliance with written procedures.

[handwritten top margin: Must be resolved by PIC in 72 hours]

What if a discrepancy is found during the monthly audit?

The discrepancy must be resolved by the PIC (or someone the PIC designates) within 72 hours of the time the discrepancy was discovered.

If the discrepancy is a theft or an unusual loss of drugs, it must be reported to the Board of Pharmacy.

Is any other monitoring required?

Yes, the devices must be inspected monthly. Inspection includes checking for:

- Proper storage
- Proper location of drugs within the device
- Expiration dates
- Security of drugs
- Validity of access codes

After the automated dispensing machines are audited and inspected, how long are the records kept?

Two years.

[handwritten: 2 years of audits & inspection]

Can the records be kept electronically?

Yes, under the following conditions:

[handwritten: Read only!]

- The records can be readily retrieved upon request.
- The records are maintained in a read-only format which does not allow alteration.
- A log is maintained showing the dates of audit and review, including which machines were reviewed, the time period reviewed, and the initials of all reviewers.

It is considered dispensing when patients are given drugs to take home with them from an emergency room. It is considered administration when a nurse or other licensed person gives the drug.

Does an automated dispensing device from an emergency room require a separate *dispensing* record?

No separate record is needed as long as:

- The automated record distinguishes dispensing from administration.
- The device records the identity of the physician who is dispensing.

[handwritten bottom margin: NO the Record distinguishes dispensing from admin.]

Under control of pharmacy

Under what conditions can a nursing home have an automated dispensing machine?

- Drugs are under the control of the pharmacy providing services to the nursing home.

- The pharmacy has on-line communication with and control of the machine.

- The nursing home has a controlled substances registration.

The doctor writes a medication order for a patient in a nursing home. Can the nurse take the drug out of the automated dispensing machine and give it?

Not until a pharmacist has reviewed the prescription and electronically authorized it. This means that the PIC must ensure there is a pharmacist with on-line access to the system available at all times to review prescriptions.

A Pharmacist must review

What are the rules for security, monitoring, and auditing of automated dispensing machines in a nursing home?

The rules are the same as in the hospital:

- Machines must be audited and inspected monthly.

- The PIC must resolve discrepancies within 72 hours.

- Drugs in the drawer must be counted when the drug is restocked.

audited/inspected monthly

A pharmacy wants to begin delivering prescriptions. Can the dispensed prescriptions be delivered to the patient's office? To the patient's home, if the patient is not at home?

Prescriptions may be:

- Directly hand delivered to patient or to the patient's agent.

- Delivered to the residence of the patient.

- Delivered to a place holding a current permit, license, or registration with the Board that authorizes the possession of controlled substances.

allowed to hold

must be delivered to residence

Can the pharmacy deliver a dispensed prescription to another pharmacy?

Yes.

Can the pharmacy deliver a dispensed prescription to the patient's doctor?

Only if the doctor is licensed to practice pharmacy or to sell controlled substances.

Can the pharmacy deliver a dispensed prescription to a person or entity that has a controlled substances registration?

Only if the person or entity is authorized to sell the drug.

What sort of written notice must be included when a dispensed prescription is delivered, instead of being picked up at the pharmacy?

A notice alerting the consumer that chemical degradation of drugs may occur under certain circumstances, and providing a toll-free or local number the consumer can call with questions.

a notice

Under what conditions can a pharmacy sell insulin?

Insulin can be sold only by a licensed pharmacist, or under the supervision of a licensed pharmacist.

only a pharmacist.

What controlled paraphernalia can only be sold by a pharmacist?

- Hypodermic needles & syringes

- Gelatin capsules

- Quinine or any of its salts in excess of ¼ ounce

What information is a person purchasing controlled paraphernalia required to produce?

- Identification, including proof of age when appropriate.

- Written, legitimate purpose for which the controlled paraphernalia is being purchased. This is not required for a "customer of known good standing."

identification – p.v.o
written legitimate purpose
known good standing

When the pharmacist sells controlled paraphernalia, what records must be kept?

- Date of dispensing

- The name and quantity of the device, item or substance

- The price at which it was sold

- The name and address of the person to whom the device, item or substance was dispensed

Date, name, quantity, price sold, name & address of the person, reason & initials of RPh

- The reason for its purchase
- The pharmacist's initials

Are there any age limits for purchasing controlled paraphernalia?

Controlled paraphernalia can't be sold to someone under 16 except by a physician or with a prescription.

Is the pharmacy responsible for safeguarding controlled paraphernalia?

Yes, the pharmacy must "exercise reasonable care in the storage, usage and disposition of such devices" to prevent diversion.

If a pharmacist violates the rules on controlled paraphernalia, what is the pharmacist guilty of?

A Class 1 misdemeanor.   *Class 1 misdemeanor*

Can SV substances be sold without a prescription?

Yes, in small quantities. The drugs that can be sold are: opium, codeine, dihydrocodeinone, ethylmorphine, and diphenoxylate.

What are the limits on each substance within a 48 hour period?

- 200 milligrams of opium
- 270 milligrams of codeine
- 130 milligrams of dihydrocodeinone
- 65 milligrams of ethylmorphine
- 32 5/10 milligrams of diphenoxylate

What are the requirements for selling a SV substance without a prescription?

- It must be dispensed by a pharmacist.
- It must be dispensed directly to the person requesting it.
- The customer must provide proof of age (18 years or older).
- The pharmacist must use professional discretion to ensure the preparation is being dispensed for medicinal use only.

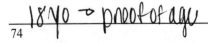 *18 y/o → proof of age*

When the pharmacy dispenses a SV preparation without a prescription, what information must be recorded?

- Date of the sale
- Name and quantity of the preparation
- Name and address of the person to whom the preparation is dispensed
- Initials of the dispensing pharmacist

Date
name
quantity

What schedules are covered by the Prescription Monitoring Program?

SII-IV.

SII -IV

When a pharmacist dispenses a drug covered by the Prescription Monitoring Program, what must be reported?

- The patient's name, address, and date of birth
- The drug and quantity dispensed
- The date of dispensing
- The prescriber and dispenser's identifier numbers
- Any other information required to comply with state and federal laws

Name, address,
date of birth,
drug & quantity
dispensed,
date of dispensing
ID #

Are there any situations where a controlled substance can be dispensed without being reported to the Prescription Monitoring Program?

Yes, reporting is not required for:

- A pharmaceutical manufacturer dispensing drugs as part of an indigent patient program
- A physician dispensing to a patient in a bona fide medical emergency or when pharmaceutical services are not available
- Administration (as opposed to dispensing)
- Dispensing within a narcotic maintenance treatment program
- Dispensing to inpatients in hospitals, hospices, or nursing homes
- A veterinarian dispensing to animals within the usual course of practice

Is it possible to apply for a waiver of all or some of the reporting requirements?

A waiver may be granted to a pharmacist with a history of compliance with the program. The waiver is granted for a special need, such as a natural disaster or other emergency beyond the control of the dispenser, or for dispensing as part of an approved research project.

If you omit information or report erroneous information to the Prescription Monitoring Program, can you be liable for civil damages?

Not if you were acting in good faith and there was no gross negligence or willful misconduct.

Is data from the Prescription Monitoring Program protected?

Yes. If you disclose data or use it for a purpose other than intended you are guilty of a Class I misdemeanor.

Class 1 Misdemeanor

Appendix – Drug Names and Schedules

"If it's stupid and it works, it's not stupid."

Throughout pharmacy school, I came up with hundreds of rhymes, jingles, acronyms, and silly memory tricks to keep drug names straight. This appendix contains the list of professional suffixes and scheduled drugs that the VA Board of Pharmacy recommends knowing for the exam. I've included any mnemonics that I created during my own studying.

For this exam, remember that most of your time should be spent learning the law. However, this appendix is provided for students who want to brush up on drug names before taking the exam.

Yes, the memory tricks are silly. But if they help you, use them!

Professional Suffixes

MD- Doctor of Medicine

OD- Doctor of Optometry

DC- Doctor of Chiropractic

DDS- Doctor of Dental Surgery

PA- Physician Assistant

RN- Registered Nurse

DPM- Doctor of Podiatric Medicine

DO- Doctor of Osteopathic Medicine

DVM- Doctor of Veterinary Medicine

DMD- Doctor of Dental Medicine

NP or LNP- Nurse Practitioner

LPN- Licensed Practical Nurse

Schedule I

Schedule I drugs are drugs which have a high potential for abuse, but which have no accepted medical use in treatment in the United Stated or which lack accepted safety for use in treatment even under medical supervision.

Schedule II

I divided the SII drugs into 3 groups and created a goofy mnemonic for each. Most of the mnemonics use the first letter or letters of the brand name.

"Mr. P. P. Toddd takes opium, codeine, and cocaine."

Brand Names	Generic Names
M.S. Contin, **R**oxanol	Morphine sulfate
Percodan, **P**ercocet, **T**ylox, **O**xyContin	Oxycodone
Dilaudid	Hydromorphone
Dolophine	Methadone
Demerol	Meperidine
	Opium
	Codeine (as a single drug entitiy)
	Cocaine

"Dex and Des met a friend to buy Ritalin."

Brand Names	Generic Names
Dexedrine	Dextroamphetamine
Desoxyn	**Met**hamphetamine
	Phenmetrazine
Ritalin	Methylphenidate

"Sue and Al subbed for Amy at the barbershop"

Brand Names	Generic Names
Sufenta	Sufentanil
Alfenta	Alfentanyl
Sublimaze	Fentanyl
Biphetamine	**Am**phetamine
Pento**barb**ital (suppositories are schedule III)	Nembutal
Seco**barb**ital (suppositories are schedule III)	Seconal
Amo**barb**ital (suppositories are schedule III)	Amytal

Schedule III

For SIII, I created mnemonics for three groups of drugs, plus some extras. Check out the Trick Questions section below for ways to remember the extras.

"2, 3, 4, pento"

Brand name(s)	Generic name
Tylenol with codeine #**2**, #**3**, #**4**; Phenaphen with codeine #**2**, #**3**, #**4**	Codeine in combination with acetaminophen
Empirin with codeine #**2**, #**3**, #**4**	Codeine in combination with aspirin
Pentothal	Thiopental sodium

"LOLcats haz TV"

Brand name(s)	Generic name
Lorcet, Lortab, **H**ycodan **A**nexsia, **Z**ydone, **T**ussionex, **V**icodin	Hydrocodone

"Mari and Ana win at bowling"

Brand name(s)	Generic name
Dronabinol	**Mari**nol
Oxandrolone	**Ana**var
Stanozolol	**Win**strol
Nandrolone	Ana**bolin**, Androlone, Deca-Dura**bolin**, Dura**bolin**, Hy**bolin**, Nandrobolic

Other SIII drugs (see Trick Questions below)

Generic name	Brand name(s)
Butabarbital	Butisol
Butalbital (unless in combination with acetaminophen, then schedule VI)	Fiorinal, Fiorinal with codeine
Thiopental sodium	Pentothal
Benzphetamine	Didrex
Phendimetrazine	Bontril, Prelu-2

Schedule IV

There are only two groups for SIV, plus some extras that will be covered in the Trick Questions section. Remember that it's not necessary to exactly match the brand name with the generic. It's enough to remember that Restoril is SIV without knowing that its generic name is temazepam.

"RAT CHALKS VXD"

Brand name(s)	Generic name
Restoril	Temazepam
Ativan	Lorazepam
Tranxene	Chlorazepate
Centrax	Prazepam
Halcyon	Triazolam
Ambien	Zolpidem
Librium	Chlordiazepoxide
Klonopin	Clonazepam
Serax	Oxazepam
Valium	Diazepam
Xanax	Alprazolam
Dalmane	Flurazepam

"Tall deer tap ten sandy ponds."

Brand name(s)	Generic name
Talwin	Pentazocine
Darvon	Propoxyphene
Tepanil, **Ten**uate	Diethylproprion
Sanorex	Mazindol
Pondimin	Fenfluramine

Other SIV drugs (see Trick Questions below)

Generic name	Brand name(s)
	Phenobarbital
Fastin, Ionamin, Adipex-P	Phentermine

Trick Questions

These questions will help you distinguish drugs that I didn't create a mnemonic for, or drugs whose schedules can easily be confused.

Which drugs become SIII instead of SII when they come in a different dosage form?

Nembutal, Amytal, and Seconal are SIII when supplied as suppositories.

Which schedules do phenmetrazine, benzphentamine, phendimetrazine, and phentermine belong to?

- *"On the 2ⁿᵈ, I met a friend"* – **phen**metrazine is SII.

- *"On the 3ʳᵈ, Bonnie and Didi had a fender bender"* – **Bon**tril (**phen**dimetrazine) and **Did**rex (**benz**phetamine) are SIII.

- *"On the 4ᵗʰ, Iona picked up the fast termite"* – **Iona**min, **Fas**tin (phen**termine**) is SIV.

Which schedules do pentobarbital, secobarbital, amobarbital, butabarbital, butalbital, and phenobarbital belong to?

- *"Pento, SECO, and amo are SECond"* – pentobarbital, secobarbital, and amobarbital are SII.

- *"B is three, but Fioricet is free"* – the drugs that start with B (butabarbital and butalbital) are SIII. However, Fioricet, which contains butalbital, is only SVI.

- *"Phenobarb is Four"* – both "phenobarbital" (which is SIV) and "four" start with an "F" sound.

Schedule V

Brand name(s)	Generic name
	Most cough syrups containing codeine
Lomotil	Diphenoxylate

Schedule VI

All prescription drugs and devices which have not been placed in another schedule are in Schedule VI. This includes any drug or device which is not in another schedule, but which is required by federal law to bear on its label one of the following legends:

1. "Rx only" or "Caution: Federal Law Prohibits Dispensing Without Prescription"

2. "Caution: Federal Law Restricts This Device To Sales By Or Use On The Order Of A Physician"

3. "Caution: Federal Law Restricts This Drug To Use By Or On The Order Of A Veterinarian"

Schedule VI also includes any drug not listed in Schedules I - V which because of toxicity, potential for harm, method of use, or collateral measures necessary to its use is not generally recognized among experts as being safe for use except by or under the supervision of a practitioner licensed to prescribe.

Brand name(s)	Generic name
Lanoxin	Digoxin
Penicillin VK	Penicillin V
Wellbutrin	Bupropion hydrochloride
Amoxil	Amoxicillin
Keflex	Cephalexin
Ultram	Tramodol hydrochloride

Made in the USA
Lexington, KY
04 June 2012